YOUR KNOWLEDGE HAS VALUE

- We will publish your bachelor's and
  master's thesis, essays and papers

- Your own eBook and book -
  sold worldwide in all relevant shops

- Earn money with each sale

Upload your text at www.GRIN.com
and publish for free

Kojo Danquah Amankrah

Assessing Poverty Situations: A Case Study of Cocoa Farming Households in the Asikuma-Odoben Brakwa District of the Central Region

GRIN Publishing

Imprint:

Copyright © 2008 GRIN Verlag GmbH
Print and binding: Books on Demand GmbH, Norderstedt Germany
ISBN: 978-3-656-87463-8

This book at GRIN:

http://www.grin.com/en/e-book/287042/assessing-poverty-situations-a-case-study-
of-cocoa-farming-households

Assessing Poverty Situations: A Case Study of Cocoa Farming Households in the Asikuma-Odoben Brakwa District of the Central Region

ABTRACT

Ghana has achieved substantial poverty reduction over the last 15 years and is on track of reducing its poverty rate by half before the target date of 2015 for the Millennium development Goals. The objective of this study is to document this remarkable achievement, and more broadly to review the evidence on a range of issues related to poverty reduction among cocoa farmers in Asikumah-Odoben-Brakwa of the Central Region using the most recent household survey data available The study aims to determine the contributions of cocoa farming activities to household income and consumption; to determine their level of access to basic services ; to determine the contribution of cocoa into the general well-being of the people and to make recommendations on how the cocoa industry could be used to improve the living standard of the Ghanaian rural communities.

The study would rely on some comparable selected Core Welfare Indicators Survey (CWIQ) for 1997 and 2003 pertaining to Asikumah-Odoben-Brakwa District.

In conclusion, the study found out that cocoa farming coupled with government provision of other social services has contributed in reducing poverty in the Asikuma-Odoben-Brakwa District thus most cocoa farming households are above the upper poverty line. Notwithstanding that, the study strongly recommends that government and other key stakeholders look at improving the health needs, provision of decent housing scheme and monitoring the usage of cocoa weighing scales to guaranty fairness.

CHAPTER ONE

INTRODUCTION

1.1 Background to the study

According to World Cocoa Foundation (WCF), cocoa is described as a smallholder crop, employing a majority of people and their families in the tropics. In the 2000 Population and Housing Census the figures revealed that cocoa farming employs over 1.5 million out of about 3,972,407 people in agricultural activities largely in rural areas of the country. According to earlier studies by NPECLC (2007), they found out that in many cases cocoa is the only source of income for these families enabling them to meet basic human needs such as food, housing, shelter and health care thus reducing poverty among most Ghanaians.

Since commercial quantities of cocoa production began in about 1879 with cocoa pods brought from Fernando Po by Tetteh Quarshie cocoa cultivation has been considered as an important economic activity. Manu.M and Tetteh E.K, (1987), however found out that, the crop is highly produced by rural farmers generally smallholders who operates family farms and cultivate acreages that range from about three acres or less in the Western North and Western South regions .The rest of the regions are Brong Ahafo, Ashanti, Central and some part of Northern Volta where rainfall is 1,000-150 millimetres per year. A few outliers may operate farms that are less than an acre, or up to about a hundred acres or more in some cases.

Tiffen, Pauline, et al (2006),cited that cocoa farming is also the basis of economic security for a wide range of people within Ghana especially women .Often cocoa farms are passed on to women as inheritances to ensure that they have an income and are protected from destitution, particularly as they age. Studies have shown that income earned from cocoa by women is more likely to be used to meet families' basic needs, such as nutrition, healthcare and education. Thus, the cocoa farming helps to reduce the poverty level among families especially in the rural areas.

Tiffen, Pauline, et al (2006), also tenant or sharecroppers frequently take up cocoa farming because it provides an opportunity for social mobility and, for some, provides enough income to send remittances to family members in other parts of the country to pay for food and children's educations. It is clear that, an improvement in the living standards of one cocoa farmer improves the lives one or two other Ghanaian.

2

Government through a number of initiatives has contributed in reducing poverty among cocoa farmers by implementing the objectives of the Ghana Poverty Reduction Programme (I and II).One of the objectives is to improve rural incomes by focusing on improving the method and quantity of cocoa produced and by increasing the farm gate price.

From 2002, government has providing insecticides and mass spraying of cocoa farms as a measure to increased output by cocoa farmers. In 2005 a consistent policy to improve agronomic practices, disease and pest control and increase value addition was pursued to increase rural incomes and to sustain the improvement of the cocoa industry. Government also continued to pay and increase cocoa bonuses to the tune of 429 billion to cocoa farmers. The Cocoa Board Scholarship Fund also increased its awards by about 25% over the previous year to the children on cocoa farmers to relieve the burden of cocoa farmers.

The COCOBOD also released 1 billion as seed money for the commencement of the proposed Cocoa Farmers' Housing Scheme.

1.2 Problem Statement.

Although most cocoa production is carried out by peasants' farmers on plots of less than three hectares a small number of farmers appear to dominate the trade. Cocoa production is also source of income for approximately 1.5 million people in production and transport. Income from cocoa farming enables families to pay for their basic commodities, medicines and schooling for their children

Indeed some studies show that about one-fourth of all cocoa farmers receive just over half of the total income. Mensah (2006), cited that one general problem identified within the cocoa households is poverty. Notably, with recent increases in cocoa prices, yields and production, one would expect poverty among cocoa farmers to have been reduced substantially over time, and indeed faster than for other population groups.

1.3 Justification

Recently, whiles much effort is being made to increase cocoa production in Ghana through pest and disease control by mass spraying exercise, paying of farmers' bonus and other incentives and the improvement of research in the area of soil fertility. Cocoa farmers in the Asikumah-Odoben-Brakwa District have not seen much improvement in their livelihoods.

This study will help key stakeholders in the industry and the authorities in the District to ascertain the impact of the cocoa farming in poverty reduction among cocoa farming household.

1.4 General Objectives

- To assess poverty incidence among cocoa farming house holds in Asikuma Odoben-Brakwa District of the Central Region.

1.4.1 Specific objectives

- To determine the contributions of cocoa farming activities to household income and consumption.

- To determine the contributions of cocoa income into the general well-being of cocoa farmers in the Asikuma -Odoben -Brakwa District.

- To determine the access of cocoa farmers to basic services such as health care, housing and nutritional needs.

CHAPTER TWO

LITERATURE REVIEW

2.0 Introduction

This chapter reviews important literature in relevant to the study. It covers topics such as measure and definitions of poverty, consequences, perspectives and concept of poverty. The chapter also covered situations of poverty in Ghana specifically national, regional and district poverty trends. The chapter also review issues relating to household characteristics of farmers in Ghana and the types of ownership of cocoa farms

2.1 Definitions of poverty and the poor

According to the World Bank (2000), "poverty is pronounced deprivation in well-being." It can also be agreed that poverty results from lack of human, physical and financial capital needed to sustain livelihoods and from inequities in access to control and benefit from reasons, be they political, social or economic.

The World Bank further described that poverty is hunger, lack of shelter, situation of being sick and not being able to see a doctor. Poverty is also not having access to school and not knowing how to read. Poverty is not having a job, is fear for the future, living one day at a time. Poverty is losing a child to illness brought about by unclean water. Poverty is powerlessness, lack of representation and freedom.

The Encarta World English Dictionary also defines poverty as the state of not having enough to take care of basic needs such as food, clothing, and housing. It also described poverty as a general word for the state of being without enough money or resources to live at a standard considered normal or basic by society.

The individual is described as poor/non-poor when consumption expenditure per day is defined to be below/above one US$ according to UN Statistical Report (2003). Thus the poor can be characterized as living in material deprivation, physical weakness: isolation, vulnerable and powerlessness.

The Human Development Report (199:15) in its report states that "a person is said to be poor if in monetary terms, 'income' or 'expenditure' falls below a level accepted as the minimum to assure a minimum descent standard of living"

According to the UN Sectary General (April 2001), he also described the poor as constituting roughly one fifth of the world's population currently living or trying to live on less than one dollar a day.

2.2 Concept and consequences of poverty

2.2.1 Concept of poverty

There are two approaches of definition to poverty that have been favoured by sociologists and researches: absolute poverty and relative poverty. The concept of absolute poverty is grounded in the idea of subsistence-the basic conditions that must be met in order to sustain a physically healthy existence. People who lack these fundamental requirements for human existence-such as sufficient food, shelter and clothing –are said to live in poverty.

The concept of absolute poverty is seen as universally applicable. It is held that standards for human subsistence are more or less the same for all people of an equivalent age and physique, regardless of where they live.

Not everyone accepts that it is possible to identify such a standard; however it is more appropriate, they argue, to use the concept of relative poverty, which relates poverty to the overall standard of living that provides in a particular society. Advocates of the concept of relative poverty hold that poverty is culturally defined and should not be measured according to some universal standard of deprivation. It is wrong to assume that human needs are everywhere identical- in fact, they differ both within and across societies.

2.2.2 Consequences of poverty

According to NPRP(1999,2000 and 2001),the major consequences of poverty include;hunger,sickness,powerless,low self-esteem,isolation,sense of helplessness, weak capacity to educate the children, inability to honour social obligations, a feeling of insecurity and vulnerability of and drudgery.

Cramer and Jensen(1994), also explained that the consequences of poverty are the inability to meet their social, economic and other household responsibilities, living in dilapidated housing, inability to afford good medical or health services with others becoming unemployed due to inadequate capital to invest in their work.

2.3 Three perspectives of poverty

According to UNDP, (HDR 1999:16) there are three approaches to poverty that has been described. The three perspectives of poverty include income, basic need and capabilities.

2.3.1 Income perspective

A person is poor if; and only if, available income is below the define poverty line. Many countries have adopted income poverty lines to monitor progress in reducing poverty incidence. Often the cut-off poverty line is defined in terms of having enough income for a specific amount food.

2.3.2 Basic need perspective

Poverty is deprivation of material requirement for minimally acceptable fulfilment of human need, including food. This concept of deprivation goes well beyond the lack of private income. It includes the need for basic health and education and essential services that have to be provided by the community to prevent people from falling into poverty. It also recognizes the need for employment and participation

2.3.3 Capabilities perspective

Poverty represents the absence of some basic capabilities to function;-a person lacking the opportunity to achieve some minimally acceptable levels of these functioning. The functioning relevant to this analysis can vary from such physical ones as being well nourished, being adequately clothed, and sheltered and avoiding preventable mobility to more complex social achievements such as partaking in life of the community.

2.4 Poverty trends

2.4.1 Poverty trend in Ghana

Ghana's average GDP per capita over the past decade was approximately US$300.The proportion of the population living in poverty fell steadily from 51 percent in 1992 to 39 percent in 2000.Though human development indicators have generally improved over the same period, a recent and past country wide surveys carried out by the Ghana Statistical Service (GSS, 2000) provide evidence of the poverty situation in Ghana since the early 1990's. Poverty levels in Ghana based on income/consumption decreased from 51.7% to 39.5% between 1991/92 and 1998/99. Extreme poverty also declined from 36.5% to 26.8% over the same period. The baseline poverty rate for 2003 was about 30.7% with extreme poverty around 20%.

Poverty incidence in Ghana as a whole had dropped to about 26.9%.However, these figures tend to camouflage the incidents of growing and deepening poverty and evidence of vulnerability and exclusion among certain groups and in some areas, especially in the Northern and the Central regions, given that on the whole population growth far outstripped the rate of decline in poverty levels during the period.

Geographical disparities in poverty tends to suggest that five(5) out of the ten(10) regions in Ghana had more than 40% of their population living in poverty in 1999.According to the recent Ghana Poverty Reduction Strategy(GPRS,2002:9),the worst affected areas are the three Northern savannah regions(i.e. the Upper East, Upper West and Northern Regions).The evidence indicates that nine (9) out of ten (10) people in the Upper East, eight (8) out of ten (10) in Upper West, seven(7) out of ten(10) in Northern region and five (5) out of ten (10) in Central and Eastern Regions were classified as poor in 1999.(GPRS,2002:GSS,2000).Of the ten regions, the Upper East, Upper West, Northern and Central regions experienced increases in poverty and extreme poverty in the 1990's.Urban areas in the Northern Savannah also experienced significant increases in poverty during the period.

The trend in the 1990's indicates varying degrees in poverty, with the largest benefits going to export farmers and wage employees and the least to food crop farmers.

2.4.2 Poverty Situation in the Central Region

The Central Region experienced increase in incidence of poverty during the 1990's.The incidence of poverty using the upper poverty line according to GSS (2000) increased from 44.3 percent in 1991/1992 to 48.4 percent contributing 11.0 percent to national poverty in 1998/1999.Using the lower poverty line, extreme poverty increased from 24 percent in 1991/1992 to 31.5 percent in 1998/1999 (GSS, 2000).

2.4.3 Selected Welfare Indicators in the Asikuma-Odoben- Brakwa

The poverty ranking incidence of the District is described as 62 per cent, 34 per cent and 35 per cent for overall, rural and urban respectively. (GPRSII 2006).The unemployed and underemployed population in the District represent 1.9 and 2.3 percent (GSS, CWIQ 2003). The adult and youth literacy rate for the district is 41.6 percent and 60.7 per cent respectively. (GSS, CWIQ 2003).Access to primary and secondary education in 2003 was 91.4 per cent and 53.3 percent respectively.

Health access and children considered to be stunted as a result of nutritional disorder is 63.4 percent and 29.6 percent (CWIQ, 2003).2.3 percent of the population have difficulty to food needs and 80.3 per cent have access to improved water sanitation. Access of the population to water, safe sanitation, improved waste disposal and access to electricity represent 98.2 percent, 40.5 percent, 96.1 percent and 37.2 per cent respectively. (GSS, CWIQ 2003)

2.5 Measuring Poverty

The strongest justification is that provided by Lanjouw and Ravallion (1996) who argues, "a credible measure of poverty can be a powerful instrument for focusing the attention of policy makers on the living conditions of the poor." The following statements are some of the reason to measure poverty

- To allow poverty issues to appear on the political and economic agenda.
- To target interventions.
- To be able to predict the effects of, and then evaluate, policies and programs designed to help the poor.
- To help evaluate institutions and understanding the politics of many government policies.

2.5.1 Steps in measuring poverty

According to Coudouel et al. (2002), there are three main steps to be taken into consideration when measuring poverty and the following steps are as follows

i) define an indicator of welfare
ii) establish a minimum acceptable standard of that indicator to separate the poor and the non-poor (often known as the poverty line) and
iii) generate a summary statistic that aggregates the information we get from looking at the distribution of the welfare indicator that we have chosen, and its position relative to minimum acceptable standards.

2.6 Dimensions of Poverty

Poverty has many dimensions. Therefore a range of indicators is needed to inform the range of policies to tackle the causes and mitigate the consequences of poverty. The analysis here therefore focuses on consumption poverty, lack of access to basic services and deprivations in human developments.

Howes, Steven and Jean Oslo Lanjouw (1997), in their studies carried out for the World Bank provided some an acceptable welfare measures to be used in Living Standard Survey and these are considered in the study.

2.6.1 Momentary Dimensions

Whilst the broader concept of well-being forms the basis of our overall approach in studying poverty, the most common approach to measuring poverty focuses on (monetary dimensions) economic welfare based on household consumption expenditure or household income, to which each resident in the household is assigned a share of the total amount.

2.6.1.1 Income

It is possible to measure household welfare by looking at household income, but there are complications that often make it a second choice for poverty analysts. People may be reluctant to report income if they have engaged in tax evasion or illegal earnings. In addition, certain types of income are not always easy to measure, such as farm income, or changes in the value of assets, such as farm inventories or the housing stock.

2.6.1.2 Consumption

Consumption measure includes both goods and services that are purchased, and those that are provided from ones' own production ("in-kind"). The consumption-based measures of living standards are generally preferred to income-based measures because it minimizes the reporting problem that is associated with the income measure of wellbeing.

If we choose to assess poverty based on household consumption or expenditure per capita, we can use an expenditure function in our analysis. Expenditure function shows the minimum expenditure required to meet a given level of utility, which is derived from a list of goods, at a given set of prices.

2.5.2 Non-momentary dimensions

There are also non-monetary measures of individual welfare, which can include indicators such as regional-level characteristics, community level characteristics and household and individual characteristics. The study would rely on some household and individual level characteristics to assess the poverty situations among cocoa farmers in the district.

2.5.2.1 Household characteristics

Some of the important household characteristics are demographic, social and economic. The demographic factors include household size, age structure, dependency ratio and gender of household head. Some of the social factors also used are health and nutritional status, education and shelter. The economic factors are employment status, hours worked and property owned.

2.6 Aggregate measures of poverty

According to Shahidur and Haughton (2007), given information on per capita consumption, and a poverty line, there are several aggregate measures of poverty that can be computed. The aggregate measures include poverty line, head count index, and poverty gap index.

2.6.1 Poverty Line

The poor are those whose expenditure (or income) falls below a poverty line. Howes, S and Lanjouw, J.O. (1997), defines poverty line as the level of consumption (or income) needed for a household to escape poverty. Poverty lines are constructed and defined based on three methods: the cost of basic needs, food energy intake, and subjective evaluations.

According to the World Bank the appropriate choice of poverty line is a matter of judgment, and will therefore vary from country to country. The poverty line for this study was considered based on the Ghana Living Standard Survey 4 report as follows: a lower poverty line of 70 Ghana cedis (about US $78) per adult per year and an upper poverty line of 90 Ghana cedis (about US$100) per adult per year.

2.6.2 Headcount index

The (headcount) poverty rate is defined as the percentage of the population (or of sub-groups within the population) with incomes below the poverty line.

The measure has several weaknesses, however: the headcount index does not capture the depth of poverty.

2.6.3 Poverty gap index

Another measure of poverty is the poverty gap index, which adds up the distances that poor people fall from the poverty line, and expresses that as a percentage of the poverty line. The poverty gap index is thought of as a way to measure the total cost of bringing each poor member of a society up to the poverty line, but it should be noted that this is not a precise measure because it requires exact information on each poor member of society, and this assumes that the government has a lot of information.

2.7 Characteristics of cocoa farming households in Ghana

The MASDAR (1998) study provides detailed description of small holder cocoa farming households and their socio-economic characteristics.

Asante (1995) suggested that farmer socioeconomic characteristics influence their managerial capacity and eventually the attainment of their household goals. The cocoa farming community was found to be an ageing one with mean age of 55(range was 30-70).Age is important as the human life cycle tends to match that of the cocoa tree itself. The apparent ageing nature of cocoa farmers is of concern as it appears to discourage replanting and suggests that the less fit has been left to carry on physically demanding cocoa maintenance practices. (Takame, 2002; Baah, 2006)

According to studies carried out by (Osei-Bonsu et al, 2001), the sex distribution among cocoa farming households suggest that the majority of cocoa farmers are men. The apparent dominance of the male in cocoa farming might be due to the relative ease with which they acquire cocoa lands and exploit their kinship ties to command labour required for cultivation and maintenance practices compared to the female(Hill,1963).
(Takame, 2002) revealed that, remittances from wards abroad are becoming significant addition to the household income.

On education of cocoa farmers, studies suggest that majority are illiterates hence their alleged 'resistance' to new technologies. Many researchers cite the reported illiteracy of farmers for the low adoption of research recommendations (Opare, 1997; Donkor et al., 1991; Boahene, 1995).

2.8 Cash sharing or remuneration of cocoa proceeds

Harris and Todaro (1970) describe three methods through which cash realized from cocoa is shared. These methods are were also highlighted by Manu and Tetteh (1987) as "abusua" and "nkotokuano".The "abunu" labourer shares the proceeds equally with landowner or his employer. The "abusa" labourer takes a third of the proceeds while the employer takes the remaining two-thirds of the proceeds. The "nkotokuano" labourer,who is responsible for plucking, fermented and drying the cocoa is otherwise referred to as immigrant-labourer.He receives a fix sum per load for the cocoa he plucks for his employer.

2.9 Cocoa mass spraying

(Amankwaa-Tia, 2008), the Ashanti Regional President of Ghana Cocoa, Coffee and Sheanut Farmers Association in addressing farmer group expressed his gratitude and appreciation to the government for embarking on the mass cocoa spraying exercise as well as providing them with inputs such as fertilizers to improve their yield.

The above statement confirms government support in ensuring that cocoa production is increased and subsequently having effect on the income of the cocoa farmers

CHAPTER THREE

METHODOLOGY

3.0 Introduction

This chapter deals with the area of study, the population and the representative sample. It further touches on the development and design of the instrument as well as the procedure to be used in collecting data for the study and how the results were analyzed.

3.1 Study area

Asikuma-Odoben-Brakwa is a tropical rainforest district with a total population of about 84,395 constituting about 5.6% of the region population. The district was curved out of the former Breman-Ajumako-Enyan-Essiam District, in 1989. Its capital is Breman Asikuma which is considered to be the least populated in the region.

The District covers a geographical area of 884.84 square kilometres. It is sandwiched between 4 main districts, at a uniform radius of about 40 kilometres apiece from each of them, except Ajumako, which is 25 kilometres away. On the northern border is the Birim South District of the Eastern border lies the Agona District and the Ajumako-Enyan-Essiam District is on the Southern border.

About 52.4% of the total population is economically engaged in Agriculture. The most important cash crop grown is cocoa. The District in 2006 and 2007 produced 100,000 and 78, 8371 bags of cocoa .As a result of its important contribution to cocoa production in the country. The following institutions Cocoa Swollen Shoot and Virus Division, COCOBOD Quality Control and the Cocoa Industry Radio Communication Network and two cocoa research station strategically located in the district.

The District lies in the semi-equatorial climatic zone, with monthly temperatures ranging between 26° -34°. Mean annual rainfall ranges between 1200 millimetres to 2000 millimetres.

3.2 Research Design

The study is a case study which was based on the response of the house head and the objectivity of the researcher.

3.3 Population

The target population was fifty cocoa farmers who are household head interviewed from five towns in the district.

3.4 Sampling Procedures

The district is divided into nine (9) cocoa growing sectors according to the (CSVDD) in the district. Out of these sectors, five sectors or towns were selected. The reason for selection was based on proximity. The farmers were selected through purposive and snow balling sampling procedure.

3.5 Instrument procedure

The researcher designed a household questionnaire as the data collecting instrument. The household questionnaire asked questions of the best-informed household head (i.e. the cocoa farmer).It was made up of 'open-ended' and 'close ended' questions. It was made of thirty eight (38) questions. The household questionnaire asked about characteristics of household and house head, educational characteristics, farming experience and acreage, access to social services, housing, assets, access to durable goods, income, health and access to government support.

3.6 Data collection

The researcher visited five (5) towns in the district assisted by the Deputy District Officer (CSVDD) in getting into the communities. The researcher used two months to collect data. The towns are Breman Amoanda, Breman Jamra, Baako, Breman Asikuma and Breman Benin.

3.7 Data analysis

The data collected was processed and analyzed. Descriptive statistics such as frequencies and percentages was used.

3.8 Sources of data

Two main sources of data were used for the study including primary and secondary sources. Primary data were obtained from the respondents and the field whiles the secondary data were collected from manuals, reports, research papers and publications.

3.49 Limitations of the study

During the study, the researcher encountered some challenges such as financial constraints, and the inaccessibility to certain towns. Another limitation was the packed scheduled of academic activities. All these limitations prevented researcher to cover a larger target population.

CHAPTER FOUR

RESULTS AND DISCUSSIONS

4.0 Introduction

This chapter explains the results by relating the data to another and the results of other workers. It also provides the implications of the results as well.

4.1 Results and Discussions

Table 1: Sex of Household Head

Response	Frequency	Percent
Male	41	82
Female	9	18
Total	50	100

Source: Field Survey, 2008

The majority of the household head from the study are males representing 82 per cent and females representing 18 percent. This confirms earlier studies carried out by (Osei-Bonsu et al 2001) that males play a dominant role in cocoa farming as the have the ability to acquire land and other resources. In addition to that male house head are considered to play a traditional role in taking care of the family. A male headed house head are considered to reduce the level of poverty in household as compared to females headed household according to World Bank Poverty Manual.

Table 2: Age of Household head

Years	Frequency	Percent
30-45	3	6
45-50	4	8
51-60	20	40
61-70	16	32
71-80	2	4
Above 80	5	10
Total	50	100

Source: Field Survey, 2008

A proportion of the cocoa farmers interviewed indicate that a large proportion falls within fifty-one to seventy years. This population can be explained as aging and thus affecting their ability to replant or increase their acreage Thus, reducing their income due to lack of strength and energy. This confirms an earlier finding carried out by Takame, (2002).

Table 3: Educational Background of Respondent

Response	Frequency	Percent
Attended School	38	76
Never Attended	12	24
Total	50	100

Source: Field Survey, 2008

The data collected should that a majority of the farmers 76 per cent have had some form of formal education that is middle school education. This implies that the majority of farmers can read and write disapproving earlier studies carried out by (Opare, 1997; Donkor et al., 1991; Boahene, 1995).stating that "majority of cocoa farmers are illiterate and this affect their ability to adapt to new technologies"

And also about 24 per cent of the household head have never attended school and had any form of formal education.

Table 4: Household size

Size	Frequency	Percent
1-3	4	8
4-5	10	20
Above 5	36	72
Total	50	100

Source: Field Survey, 2008

Most of the household have a large number of members above five representing about 72 per cent and reflecting a possible correlation between the level poverty and household composition. Obviously, number of household members or size influences the per capita income of the family meaning that the house head need additional income to cater for the needs a larger house hold size above three as compare to average household size of three.

This could explain why a proportion of the farmers say that their income is low and inadequate.

Table 5: Cocoa Farming Experience

Years	Frequency	Percent
1-5	6	12
6-10	3	6
11-15	8	16
16-20	18	36
Over 20	15	30
Total	50	100

Source: Field Survey, 2008

The table indicates that a majority of the farmers have been cultivating cocoa for over ten years. Thus cocoa cultivation is important occupation as main source of income for the poor and thus contributing to poverty reduction confirming studies carried out by Harold Coulombe and Quentin Wodon (2007).

Table 6: Secondary Occupation of Respondents

	Frequency	Percent
Other crop farming	24	48
None	19	38
Other occupation	7	14
Total	50	100

Source: Field Survey, 2008

The table above indicates that, about 48 per cent of the cocoa farmers indicate that they are engaged in other crop farming such as oil palm plantation, citrus and cereals cultivation as means to supplement their major income from cocoa cultivation.

Also, about 38 per cent indicated that they do not have any secondary occupation as a source of supplementary income thus cocoa cultivation is an important source of income to their families.14 per cent of the cocoa farmers indicate that they are engaged in other economic activities such as sale of cement and petty trades to supplement their income from cocoa cultivation.

Table 7: Employment Status of household members

Occupation	Frequency	Percent
Employed	15	30
Unemployed	35	70
Total	50	100

Source: Field Survey, 2008

The table reveals that, about 70 per cent of the households interviewed showed that they do not have any members of their household engaged in ay form of employment. The high unemployment status of household members indicates a higher dependency ratio. Thus, making the income from cocoa not sufficient to cater for the basic needs of household members and increasing the household susceptibility to poverty.

Also 30 per cent of the households have members who are employed thus reducing the dependency ratio.

Table 8: Type of Cocoa farm Operated

Type	Frequency	Percent
Owner	30	60
Abusua	5	10
Abunu	15	30
Total	50	100

Source: Field Survey, 2008

Majority of the farmers constituting about 60 per cent of the total farmers own their farmers implying that they have absolute share of their incomes received from cocoa cultivation.30 per cent of the farmers responded that they share their income equally with other party whiles 10 per cent of farmers cultivate the cocoa for the family from which the family head is given his part of the income.

Table 9: Size of Acreage

Acreage	Frequency	Percent
0.5-2.4	5	10
2.5-3.4	4	8
3.5-4.4	5	10
4.5-5.4	14	28
5.5-6.4	10	20
6.5-7.4	6	12
Above 7.5	6	12
Total	50	100

Source: Field Survey, 2008

About 28 per cent of the farmers owns between 4.5 to 5.4 acreage of cocoa farms constituting the highest whiles the lowest is 8 per cent cultivating cocoa farms between 2.5 to 3.4 acres. The results from the table thus confirm a statement made by the World Cocoa Foundation that cocoa farmers are smallholder farmers found in the tropics.

Table 10: Proportion of income Received by owner farmers

Income GH	Frequency	Percent
<100	1	3.3
100-499	12	40
500-899	10	33.3
900-1299	3	10
1300-1699	2	6.67
1700-2099	1	3.3
2100-2599	2	6.67
Total	30	100

Source: Field Survey, 2008

About 40 per cent of owner farmers earn an income between 100-499GH¢ from cocoa cultivation representing the highest proportion whiles 3.3 per cent of the farmers receives lowest proportion of income below 100 GH ¢.The income thus receive is above the poverty line using the head count index of poverty measure.

Table 11: Proportion of Income received by Abunu farmers

Income	Frequency	Percent
100-499	10	66.67
500-899	5	33.33
Total	15	100

Source: Field Survey, 2008

The results from the table, explains that about 66.67 per cent and 33.33 per cent per cent of "abunu" farmers earns income between 100-499 and 500-899 GH ¢ respectively.

Table 12: Income from Cocoa Production within the last growing season

Income GH¢	Frequency	Percent
<100	2	4
100-499	21	42
500-899	15	30
900-1299	4	8
1300-1699	3	6
1700-2099	2	4
2100-2599	3	6
Total	50	100

Source: Field Survey, 2008

From the table, 42 per cent of the farmers received an income between 100 to 499 GH¢ represent the highest proportion whiles 4 per cent of the farmers the lowest income below 100GH¢.In general the income received farmers irrespective of the ownership in the last growing season is above 100 GH ¢ which indicates that the poverty status of the farmers is above the lower poverty line of 70 GH ¢ and above upper poverty line of 90 GH¢.

From the above results, it can be said the cocoa farmers are not poor using the headcount index measure of analyzing poverty.

Table 13: Farmers Subjective Perception on Income from Cocoa

Perception	Frequency	Percent
Very low	7	14
Low	19	38
Sufficient	24	48
Total	50	100

Source: Field Survey, 2008

From the results obtained, about 48 per cent of the farmers indicated that their income earn from cocoa cultivation is sufficient and 52 per cent also indicated that their income from cocoa cultivation is very low and not adequate to cater for their basic needs.

Table 14: Income satisfaction from cocoa bonus

Satisfaction	Frequency	Percent
Satisfactory	31	62
Very bad	14	28
None	5	10
Total	50	100

Source: Field Survey, 2008

From the table, 62 per cent of the farmers indicated that the income received from the cocoa bonus is satisfactory as it supplement their income needs whiles 28 per cent of the responded indicated that the income from cocoa is very bad and not able to meet some of their basic needs.10 per cent could not express their satisfaction about their income.

Table 15: Benefit from Mass spraying in the last growing season

Response	Frequency	Percent
Yes	40	80
No	10	20
Total	50	100

Source: Field Survey, 2008

From the results obtained, 80 per cent responded that their farms have been sprayed in the last growing season whiles 20 per cent responded that their farms have not been sprayed.The other 20 per cent attributed their inability to benefit from the mass spraying mainly to political and lack of information about the exercise.

The mass spraying is beneficial as it reduces the cost of production thus increasing the income of the cocoa farmers.

Table 16: Access to durable goods as a result of cocoa farming

Goods	Frequency		Total Frequency	Percent	
	Yes	No		Yes	No
Radio	38	12	50	76	24
Bicycle	23	27	50	46	54
Sewing Machine	30	20	50	60	40
Mobile phone	31	19	50	62	38
Television set	33	17	50	66	34
Fridge	20	30	50	40	60
Motorbike	13	37	50	26	74
Vehicle(private)	16	34	50	32	68
Knapsack	15	35	50	30	70
Rechargeable lamp	21	29	50	42	58

Source: Field Survey, 2008

Radio, sewing machine, mobile phones, and television set were among the most durable goods that were acquired by majority cocoa farmers. Most of the farmers see the above goods as a necessity as compared to the less acquisition of bicycle, fridge, motorbike, vehicle, knapsack and rechargeable lamp. Their inability to acquire most durable goods could be related to the fact the household do not have enough to spend on durable goods as their basic needs such as food take much of their income.

Table 17: Acquisition of assets as result of cocoa farming

Assets	Frequency		Total Frequency	Percent	
	Yes	No		Yes	No
Land	45	5	50	90	10
House	41	9	50	82	18
Savings	34	16	50	68	32

Source: Field Survey, 2008

Majority of the cocoa farmers representing 90, 82 and 68 per cent responded that they have been able to acquired land, house and savings respectively as a form of assets as a result of income received from cocoa cultivation.

Table 18: Household members' coverage under NHIS

Response	Frequency	Percent
Yes	36	72
No	14	28
Total	50	100

Source: Field Survey, 2008

From the table, 72 per cent responded that they have been able to register their household members under the National Health Insurance Scheme whiles 28 per cent responded they have not register under the scheme because their incomes are not adequate to cover household members under the scheme.

Table 19: Type of Housing

Housing	Frequency	Percent
Mud	34	68
Cement	16	32
Total	50	100

Source: Field Survey, 2008

From the table, 68 per cent live in mud houses whiles 32 per cent live in the houses built with cement. This could imply that the income received from the cocoa farming activities is not adequate enough to use on improving their housing conditions after expending on their basic needs. The study also showed that about 20 per cent of the households have toilet facility in their homes whiles 80 per cent use the KVIP.

The study also revealed that majority of household members 98.5 per cent have access to community clinics, access to community bore whole, access to electricity and the ability to send their school age child to school. This findings is confirmed by an earlier study conducted by Coulombe .H and Wodon.Q. (2007),which stated that the ability for the rural dwellers to have access to the above basic services is due to strategies outlined in GPRS document in reducing poverty in the rural areas.

From the study, about 98 per cent of the households have enough income from cocoa cultivation to meet their food needs especially children food three times a day. The ability of cocoa farming households to have enough income for food needs shows that the family is above the cut-off poverty line. This is in line with an earlier statement made by Shahidur and Houghton (2007).

CHAPTER FIVE

SUMMARY, CONCLUSIONS AND RECOMMENDATIONS

5.1 Summary

The findings provided by this study showed that cocoa farming is an important activity in improving the living standards of cocoa farming households. Cocoa farming is considered as a primary source of income to the households as they are able to feed their families, acquire assets and durable goods.

It can also be said that, cocoa farming contributes meaningfully to the general well being of cocoa farming households as it correlates positive with the household income and consumption levels. The findings also showed that government role in providing infrastructure and services such as the National Health Insurance Scheme, provision of community bore holes and clinics and the mass spraying exercise go along way in reducing the poverty among cocoa farming households.

In conclusion the study found out that, cocoa farming coupled with government provision of social services has contributed in reducing poverty in the Asikuma-Odoben-Brakwa .In a nut shell the poverty situations among the cocoa farming households is above the upper poverty line.

5.2 Conclusions

- Though a majority of farmers consider their income from cocoa cultivation as inadequate.Thus, the combinations of income from sale of the beans, cocoa bonus and the engagements of majority of the respondents in other economic activities mostly crop farming constitutes a major source of income to cocoa farming households.

- A large portion of cocoa income is spend on feeding of household members as it is considered as an important function of consumption pattern of cocoa farming households. In addition to that the income from cocoa cultivation has enable cocoa farming households to acquire assets such as land and house and some other durable goods such as radio, sewing machine, mobile phones for cocoa farming households.

- The study also revealed that majority of the cocoa farmers do not live in a decent housing.

- The study also revealed that the mass cocoa spraying exercise by government is an appropriate step increasing yield of the crop though a few cocoa farmers complain of politicizing of the exercise.
- The study revealed a number of cocoa farming households are not covered under the NHIS due to financial constraints.

5.3 Recommendations

1. Cocoa farmers should develop their entrepreneurial ability by engaging in other economic activities so as to supplement their income.
2. Government and the private sector should explore the economic potentials of the district so as create jobs for households to reduce the dependency ration on household income.
3. The District Assembly should collaborate with Local Traditional Authority so as to review the land tenure system in Baako and Jamera to enable cocoa farmers to have fair share of their income received from cocoa production.
4. Government in partnership with COCOBOD or non-governmental organizations such as Habitat for Humanity should develop affordable housing scheme or rehabilitate cocoa farmers' dilapidated buildings.
5. The District Directorate of Agriculture should monitor the activities of the extension officers to render regular services to cocoa farmers.

45.4 Suggestions for further studies

I recommend that further studies should be conducted in the use of weighing scales in measuring the cocoa beans as most farmers complain of unfair appropriation of scales

REFERENCES

Amankwaa-Tia, N.K. (2008). "Chief Farmer commends government for increasing producer price of cocoa." *The Ghanaian Times*, Friday, March 7, pp25

Asante, E.G. (1995). *An empirical analysis of the impact of farmer managerial capacity on technology adoption in a developing market economy: a case study of cocoa farming in Ghana.* Report submitted to Winrock International, Arlington, and U.S.A

Baah, F. (2006). *Cocoa cultivation in Ghanaian analysis of farmers' information and knowledge systems and attitudes.* Ph.D thesis, Institute of International Development and Applied Economics, University of Reading, Reading, U.K, 292 PP

Boahene, K. (1995). *Innovation adoption as a socioeconomic process: The case study of Ghanaian cocoa industry.* Amsterdam: Thesis Publishers.

Coudouel et al. (2002*). Poverty Measurement and Analysis.* PRSP Sourcebook, World Bank, Washington D.C.

Coulombe, H. and Wodon.Q. (2007).*Poverty, livelihoods, and Access to Basic Services in Ghana.* Ghana CEM: Meeting the Challenge of Accelerated and Shared Growth.

Cramer, L.G. and Jensen, C.W. (1994). *Agric Economic and Agribusiness.* John Wiley and Sons.Inc 6th edition, Canada

Donkor, M.A., Jones, A.P and Henderson, C.P. (1991).*Survey to quantify adoption of cocoa production technologies in Ghana.*FSU research report 3, Tafo: Cocoa Research Institute of Ghana.

Ghana Statistical Service, (1995). *The Pattern of Poverty in Ghana, 1988-1992.* Accra: January.

Ghana Statistical Service, (2003). *Core Welfare Indicators Questionnaire Survey 2003 Main Report* March.

Ghana Statistical Service, (1999). *Ghana Living Standard Survey 4 Data Users Guide.*

Ghana, Statistical Service, (2000). *Poverty Trends in Ghana in the 1990s.* October.

Ghana, Statistical Service, (2000). *Ghana Living Standards Survey*: Report of the Fourth Round October.

Harris, J.R and Todaro, M.P. (1970). *Migration, Unemployment and Development Sector Analysis,* American Economic Review (p.60)

Howes, S and Lanjouw, J.O. (1997). *Poverty Comparisons and Household Survey Design,* LSMS Working Paper no. 29. The World Bank

Lanjouw, J.O and Ravallion (1996). "How Should we Assess Poverty Using Data from Different Surveys." From *Poverty Lines Newsletter September.* The World Bank

Manu.M and Tetteh E.K, (1987). *A Guide to Cocoa Cultivation.* Published by Cocoa Research Institute of Ghana

MASDAR (1998). *Socio-economic study of cocoa farming communities in Ghana.* Accra: MASDAR International consultants/Ghana Cocoa Board.

Mensah, P. (2006) *The Socio-economic impact of the cocoa industry on the farmer: A case study of the Ejisu-Juabeng District of the Ashanti Region.* Unpublished B.Sc Dissertation submitted to the Department of Economics, University of Cape Coast, and Cape Coast.

Newton N. et al. (1999). *Encarta World Dictionary.* Bloomsbury Publishing Plc, First Edition.

NPECLC (2007) *Pilot labour survey in cocoa production in Ghana, 2006.*Accra: Ministry of Manpower, Youth and Employment/National Programme for the Elimination of Worst Forms of Child Labour in Cocoa (NPECLC).

Opare, K.D. (1997). "The Role of Agricultural extension in the adoption of innovations by cocoa growers in Ghana." *Rural Sociology,* 42(1):72-82.

Osei-Bonsu,K.,Baah,F and Afrifa,A.A.(2001).*Evaluation of cocoa conditions, problems and possible solutions in four farming communities surrounding the Kakum forest reserve, Central Region,Ghana.*Consulting report for Conservation International,Tafo:Cocoa Research Institute of Ghana.

Shahidur, R.K and Haughton (2007). *Poverty Manual.* Published World Bank Institute, JH Revision

Takame, T (2002). *The cocoa farmers of southern Ghana: incentives, institutions and changes in rural West Africa.* Institute of Developing Economies, Japan Trade Organisation, Occasional papers series no.37.

Tiffen, Pauline, et al (2006). *From Tree-minders to Global Players: Cocoa farmers in Ghana in Chains of Fortune*: Linking Women Producers and Workers with Global Markets. p. 12. Commonwealth Secretariat. London, UK

UNDP *Human Development Report (1999:15)*

United Nations, (2003) *Statistical Report*, New York

UN Secretary General (April 200), *A chance for the world's poorest* by the United Nations

Web resource: http://go.worldbank.org/0C60K5UK40

Web resource: http://worldcocoafoundation.org/

Appendix: Questionnaire

University of Cape Cost

School of Agriculture

The information collected in this questionnaire solely for academic purposes and the confidentiality of the respondents shall be protected.

Locality Name.......................   House No...............................

Characteristics of Household

1. Name of Household Head.....................................

2. Sex   a. Female [   ]   b. Male [   ]

3. Age of Household Head   a.30-45 [   ]   b. 45-50 [   ]   c. 51-60 [   ]
d. 61-70 [   ]   e. 71-80 [   ]   f. above 80 [   ]

4. Marital Status   a.   Married [   ]   b. Single [   ]   c. Divorced [   ]

   d. Never Married [   ]

5. Have you ever attended school?   a. Yes [   ]   b. No [   ]

6. If yes, what is your highest level of schooling?

a. Nursery/Kg [   ]   b.Primary [   ]   c.J.S.S/S.S.S/Middle School [   ]

d. Vocational [   ]   e.Other [   ] ....

7. What is the Household Size?   a. 1-3[   ]   b. 4-5 [   ]   c. Above 5 [   ]

8. How many children are in the household?.....................

9. What is type of cocoa farm you operating? a. Owner Operated [   ]

c. Abusua Type [   ] d.Abunu Type [   ]   d. Other [   ]..................

10. How long have you been a cocoa farmer? a. 1-5yrs [   ]   b. 6-10yrs [   ]   c.
11-15yrs [   ]   d. 16-20yrs [   ]   e.Over 20yrs [   ]

32

11. The Secondary Occupation of household head if any?

a. Other Crop Farming   [   ]   c. None [   ]

b. Other Occupation (specify) [   ]   …………………………………..

12. What is the size of cultivating acres? a.1 acre [   ] b.2 acres [   ] c.3 acres [   ]

 d.4 acres [   ]   e.Other [   ]   …………………………..

13. How many of the household are currently working?…………………………

Social Status Indicators

Housing

14. What is your house made of?

a. Wall

Please Indicate (1, 2, 3...) [   ]

i.Cement….1   ii.Bricks………2   iii.Wood……..3   iv. Thatch……4

v. Mud………5   vi. Other (specify)…..6

b. Roof

Please Indicate (1, 2, 3...) [   ]
i.Thatch….1   ii. Asbestos…2   iii. Concrete… 3   iv. Wood….4

v. Galvanized sheets…5   vi. Brick tiles….6   vii.Other (specify)…7…………….
c.Floor

Please Indicate (1, 2, 3...) [   ]

i. Cement/tiles…1   ii. Wood…2   iii.Mud…3   iv. Earth floor...4   v.
Other (specify)….5……………….

13. Do you have toilet facility in this house?

a. Yes [   ]   b. NO [   ]

15. What is the main source of drinking water supply to this household?

Please Indicate (1, 2, 3...) [   ]

I. Pipe-borne water…..1   ii. Bore hole/well ….2   iii. Rain water collection…….3

iv. Surface water (river, stream, dam, lake)…..4   v. Other………….5

16. How long does it take you to the main source of drinking water?

a. 10-20min [   ]   b.21-30min [   ] c.31-45min [   ] d. above 45min [   ]

17. Do you have electricity?

a. Yes [   ]   b. No [   ]

18. What is the main source of energy used in cooking?

a. Firewood [   ]   b.   Electric Stove [   ]

c. Gas   [   ]   d.   Other (specify) [   ] ………………

Access to Durable Goods

19. Do you have the following in the household?

GOODS	a.YES-1 b.No-2
a. Radio	
b.Bicycle	
c.Sewing Machine	
d.Mobile phone	
e.TV	
f.Fridge	
g.Motorbike	
h.Vehicle	
I. Motorized mist blower/knapsack	
j.Rechargeable lamp/Charging battery	

Income

20. How many bags of cocoa did you produce in the last cocoa season?.................

21. How would you consider the output of to the last cocoa season to the previous season?

a. High [   ]   b. Low [   ]

22. How much did you earn an average from cocoa farming activities in the past 12 months?...............

23. Do you have someone outside this household who sends money to any member of the household?

a. Yes [   ]   b. No [   ]

24. In the past 12 months, how much money in total has he/she sent to any member of the household?...............................

25. How will you consider your income from cocoa production?

a. Very low [   ]   b.Rather low [   ]   c. sufficient [   ]   d Rather high [   ]

26. How will you describe the payment of bonus from cultivation of cocoa?

a. Very Satisfactory [   ]   b.Satisfactory [   ]   c. Very Bad [   ]

d. Bad [   ]   e.None [   ]

27. Are you able to provide all the required three square meals for the members of the household?

a. Yes [   ]   b. No [   ]

Health Status

28. What are the main medical care options available to members of the household who fall sick or injured?

Health facility	Tick as appropriate
a. Clinic/Hospital	
b.Buy drugs from pharmacy/chemical shop	
c.Care from traditional healer	
d.Treat with any tablet/herbal medicine	
e.Other (specify)………………………………………………	

29. Are all members in the household covered under the National Health Scheme?

a. Yes [   ]   b.No [   ]

30. If no, what are the reasons?.....................................

31. What is the distance from your house to the nearest medical/health centre?

a. Less than a mile [   ] b. 2-3 miles [   ]   c. 3-4 miles [   ] d.5 or more miles [   ]

32. Within the last 12 months did any member of the household suffer from the following diseases? Tick as appropriate

a. Diarrhoea [   ]   b. Malaria [   ]   c. Cough [   ]   d. Measles [   ]

e. Kwashiorkor [   ]   f. Other (specify) [   ]....................

Educational Status of household

33. What is the highest level of education for household members?

a. Nursery/Kg [   ]   b.Primary   [   ]   c.J.H.S/S.H.S [ ] d.Vocational [ ]   e.Other [   ]

34. Are all members of the household of school going age currently in school?

Yes [   ]   NO [   ]

35. If no, what are the reasons?

a. Unable to afford fees/educational materials [   ]   b. Children needed as labour on farm [
]   c.Unavailability of schools in
the community [ ] d. Other [ ]....................................

36. Has any of your children (qualified) in the SHS level awarded with COCOBOD
Scholarship? Yes [   ]   No [   ]

Others

37. How often did an extension officer visit you in the last season?

a. Once a year [   ]   b.Once every 3 months [   ] c. Irregular [   ] d. None [   ]

38. How will consider their services?

a. Very Satisfactory [   ]   b.Satisfactory [   ]   c. Very Bad [   ]

d. Bad [   ]   e. None [   ]

39. In the last 12 months did you benefit from government support in the following?

Extension Service	Yes—1	No -2
Mass spraying		
Insecticides		
Herbicides		
Fertilizers		
Purchasing of Seedlings		
Farm Equipment		
Other(please specify)